Mrs Light and Mr Dark

MIKE PEARCE

MIKE PEARCE

Copyright 2018 by Mike Pearce

All rights reserved. No part of this book may be reproduced, distributed or transmitted in any form or by any means, including photocopying, recording, or other electronic or mechanical methods, without the prior written permission of the author, except in the case of brief quotations embodied in reviews and certain other non-commercial uses permitted by copyright law. You must not circulate this book in any format.

This book may not be resold or given away to other people. Please respect the work of the author and purchase a copy for you own use.

This is a fictional work and all characters are drawn from the author's imagination. Any resemblance or similarities to persons living or dead are entirely coincidental

Copyright © 2018 Mike Pearce

All rights reserved.

ISBN:10-1725176122
ISBN-13:978-1725176126

DEDICATION

This book is dedicated to all those who enjoy the light and
those night owls that work at night

CONTENTS

Acknowledgment
Preface

1	The couple	1
2	The world of Mr Dark	3
3	The world of Mrs Light	5
4	Mr Dark takes over for now	7

ACKNOWLEDGMENTS

The author would like to thank Christine Pearce for reading and checking through the manuscript.

MIKE PEARCE

PREFACE

All of us depend on Mrs Light for our existence. Mr Dark does not get a look in, although he was more common in the past. This story looks at the constant relationship of Mrs Light and Mr Dark and what happened when Mrs Light left for good.

MRS LIGHT AND MR DARK

John Oxenham 1913 from Bees in Amber

Where is Darkness more deadly than death itself

There is Blindness beyond that of sight;

There are souls fast bound in depths profound of unconscious and heedless Night.

William Shakespeare. Twelfth Night

Act 4 Scene 2 (Feste)

You're wrong madman. There is no darkness except ignorance, and you're more ignorant than a plague of fog

1 THE COUPLE

Mrs Light was most happy when there were no clouds in the sky and the sun was shining. She also loved situations at night where there were other sources of light so she could show her face again.

Mr Dark on the other hand loved blackness and any place where no light and shadows were formed, especially in the darkest places.

Often Mrs Light was separate from Mr Dark, but nearing dusk they overlapped and were able to talk to one another. Sometimes Mrs Light would continue to be present in a place on earth with no sunset for several months or if there was another source of light in the dark. Mr Dark could also do the same but this was in the extreme north. As you move away from the pole the number of days with sunlight or 24 hours of night decreases.

With Mrs Light you always knew what was happening but with Mr Dark hidden or sudden things would often occur, without anyone knowing, Mrs Light was always associated with brightness. Brilliant, cheerful, vivid colours were always seen clearly in the light but not in the dark. Even a black cat looked grey in the dark.

2 THE WORLD OF MR DARK

This rayless, pitch-black world was often unpromising, undiscerning, vague, viscious and without hope.

Many people have their own darkness, living in dark, dim, murky, dingy rooms, miserable, melancholy and desolate, Goths see beauty in everything dark and often wear black like dark music and dark characters. Mystery surrounds Mr Dark and his darkness can contain the invisible and the impassable like in caves underground. Turn on a light and the caves may be beautiful, full of crystalline structures and features. One may grope in the dark along this secret way of passages concealed along its route which may lead to nowhere. Often Mrs Light will try to creep as in the checkered pattern on the ground between leaves.

Mr Dark has not been associated with good. 'The subjects of the kingdom will be thrown outside, into the darkness, where there will be weeping and gnashing of teeth' (Matthew 12;8 NIV.)

Mr Dark is often associated with evil; a foreboding, works of great darkness, sorcery, princes of darkness and spells. Who knows what lies down in the dark alley or cellar or coal mine waiting in the shadows? The children's verse – 'In a dark, dark house there was a dark, dark room and inside the dark, room there was a dark, dark corner there was a big black box and in the big black box there was a------------- ghost' is used emphasise the dangers of darkness and the unknown.

What lurks in the deep, bubbling, sulphuric muddy waters of the bowl of a volcano or the bottom of the dark, deep oceans? Mr Dark can see all the dark deeds done at night; murders, robberies, abuse and vandalism. However the influence of darkness can be overcome by those people who are blind and whose other senses are heightened.

3 THE WORLD OF MRS LIGHT

When Mrs Light was at her best everything looked clear, distinct and visible. Even the sea or other areas of water would shine and sparkle like the stars on the ripples. Her light allowed one to see beneath the waters to see life. Mrs Light would always try and sneak into Mrs Darks's world, even in the darkest forest. Light would creep though the canopy of trees. With Mrs Light People are encouraged to be happy, joyful and hopeful.

Mrs Light is often associated with good- Jesus, the Light of the World.

A white object will reflect a lot of Light, there is no shadow when light is directly from above. Green and blue pixels on a screen illuminated appear white and when not illuminated look black. Your pupil in your eye will dilate to allow as much light in as possible.

Mrs Light does not have to be from the sun. A glow worm or other forms of chemical luminescence or phosphorescence can break the effect of Mr Dark. Mrs Light gets her chance at firework displays where she exhibits herself in all colours and exploding forms, much to the enjoyment of onlookers.

To counter Mr Dark's influence the Prescription Act in 1832 gave homeowners the right to Mrs Light if they had had access to it for twenty years. No one can erect a building, a wall or even plant trees that would reduce the amount of light entering a building through a window etc. Mr Dark gets around this if someone decides to enlarge their window or put in a new window they have to wait another twenty years for the act to come into force. Often in alleys or older buildings one may see the signs 'Ancient Lights' under windows. However not all Mrs Light's actions can be praised. In the visible range of the spectrum much of her influence is good. Even outside this, some of her influence can be useful as in X-rays, UV/gamma wave treatments/diagnoses etc. Unfortunately her usefulness may be exploited in the radiation/blasts received from bombs especially atom bombs where thousands of people can be killed.

4 MR DARK TAKES OVER FOR NOW

Mixing paints primary or secondary colours, together always gives a darker colour in paintings. Darkness help to bring out the light. Mr Dark could be more deadly than death itself. Many people remained ignorant of what happened with Mr Dark. His world in the beginning was inky black even for the lonely. He was really upset when he was left out of the spectrum and the rainbows.

Mrs Light had always had an advantage over Mr Dark, the growth of plants animals and people and society itself have depended on her for thousands of years, but the time had come for Mr Dark to put an end to all this. The earth lay fairly dormant for millions of years except for the odd earthquake or volcanic eruption. However, in the past, as depicted by fossil evidence, near total destruction of life occurred several times. Mr Dark knew something again was going to happen. He said to his wife, Mrs Light, he would not see her again for a very, very long time. Mrs Light was upset and wondered what was the reason for him leaving. It was not long before she understood. She felt herself being propelled from the centre of the earth to the surface illuminating the whole sky with reds and yellows and blues. With this epic surge high up into the atmosphere went molten rock and enormous superheated clouds of ash and dust. This continued for days and days, the dust and debris gradually falling from the sky all around the world under gravities influence and other huge volcanoes were triggered by this catastrophic event. Gradually the blue sky disappeared worldwide, the patches remaining getting smaller and smaller until eventually they were non-existent.

Solar power on earth had fueled crops which grew for a week or two but without sunlight the leaves turned brown and perished. It was very hot, little heat being able to escape into the atmosphere. People tried to leave by all means of transport hoping that in other parts of the world Mrs Light would still be there, but she wasn't. People relied on stored food, frozen and canned, but these soon ran out and anarchy erupted.

Gangs fought gangs as in primitive times to raid food until they also finally succumbed to the inevitable that all life would die. The only life flourishing was in the Atlantic trenches, the animals taking their nutrients from volcanic chimneys deep in the oceans. Looking across the earth it was like a graveyard, the buildings still standing talk and erect. There was no traffic noise not even a bird singing. The only noise heard was the wind or the clatter of signs or ropes on the masts of empty boats. The vermin had survived, feeding on anything dead but even they were short of food now.

Mrs Light waited and waited but the earth remained dark except for the odd flash of lightning emanating from the dark grey clouds. Eventually after millions of years the core of the earth settled and volcanic activity was greatly reduced and gravity, together with rainfall, had played its part in bringing down most of the dust particles suspended in the air. Mrs Light crept in past her husband, firstly in short bursts and eventually she started to share some of the day again.

Seeds germinated, and new plants and trees emerged and eventually something stirred on the earth to become dominant. This time it was not dinosaurs or man. Something both black and white, a new offspring from Mr Dark and Mrs Light.

Leonard Bernstein and Stephen Sondheim

West Side Story. Tonight (Quintet). Tony

Today, the minutes seem like hours

The hours go so slowly

And still the sky is light

Oh moon grow bright

And make this endless day endless night

To see other publications below by the author visit
snappysnappybooks.com

The really, really, really useful series

How to be a Successful Business Weed
How to Deal with Life's Snakes and Ladders
Know Your Students and Build Your Image
Pens for Pops
How to be a Successful Charity Shop
Make up-revealed
Ronnie's Sermon snippets
Wastefulness-Bone and Urine
Fertility Stones and Chocolate Eggs
Clingers, creepers and scramblers
I Herring Gull
Viking Bay-Natural History
Go Fat Go
Hidden from the heart but not forgotten
Pulvi Royal

Other books by Mike Pearce:

Pattern for Purpose- God's and Man's designs
Red Fred Cell and Friends
Human Termites eat London
Pigeons Splat London
Glass Anemones Tentacle-ize London
Tuppeny Hangover
I am Termite
The littlest Oyster
Bits and Bobs
The Shell Man
Cats at Christmas
Tails, Tales
Trust-Nothing but a Must
In a Dark, Dark Corner was the Holy Ghost
The Shell Lady
Captain Grottbuster versus the Grey World
London's Nemesis (Trilogy of 3, 4 and 5 above)
Saved by Angels (Trilogy of 6, 8 and 14 above)
The World of Wax
Photosynthetic Women
Queen Rat on Deadman's Island
The Watcher on the Fal
The Rock Pool
The Little Shepherd Boy's Gift
The Living Fossils
Old Mother Nature Laughed and Laughed

MRS LIGHT AND MR DARK

Betty's Barcodes
Time Runs Dry (play)
Valentines Cards
The Scrofula Infirmary
The Cornish Urchin
My Therizinosaurus
Spider in the Tomb
The White Cockerel
The Red Church Doll
Butterfly Angels (compilation of previous books)
The Girl Under the Paeony Tree
Baby Feet
The Sparrows' Last Soul
Ball Rooms
Absorbed by a Woman
St Mildred-Patron Saint of Thanet
The Slothful Wife
The Tuppeny Bear
The Boy who found Christmas
Nothing but leaves
The Giant's Toothpick
The Night Mare
The Old Pot and the Golden Shoes
Sitting next to Angels
Exodus to a leaf
The forlorn fruit fly
The Pawnbroker's Souls
The Nursery Rhyme Cat
A Call Under the Sea
Dead Donkey Lane
A Googolplex of Mice
The Eggstraordinary Easter Egg
The China Blackbird
A slice of Slang with a touch of Cockney and a

drop of Dorset
The Rusty Gate
Beware of Cucumbers, apples and pigs
The Lady loves Red
The woman who smelled books
Till my lips were salt as brine
The Giant and the Giraffe Boy
The man who always sprinted
Boy, could she smell!
Coloured bricks
The Lady who loved Hairspray
I'm just going to the bathroom

ABOUT THE AUTHOR

Dr Mike Pearce is a scientist interested in behaviour. He also was a lecturer in human biology and health at a college in Canterbury, Kent

www.ingramcontent.com/pod-product-compliance
Lightning Source LLC
Chambersburg PA
CBHW030131230526
45469CB00005B/1899